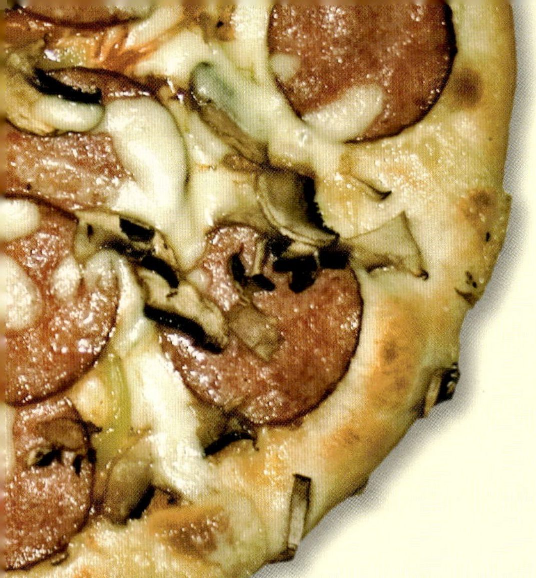

Pizza Party

by Allison K. Lim

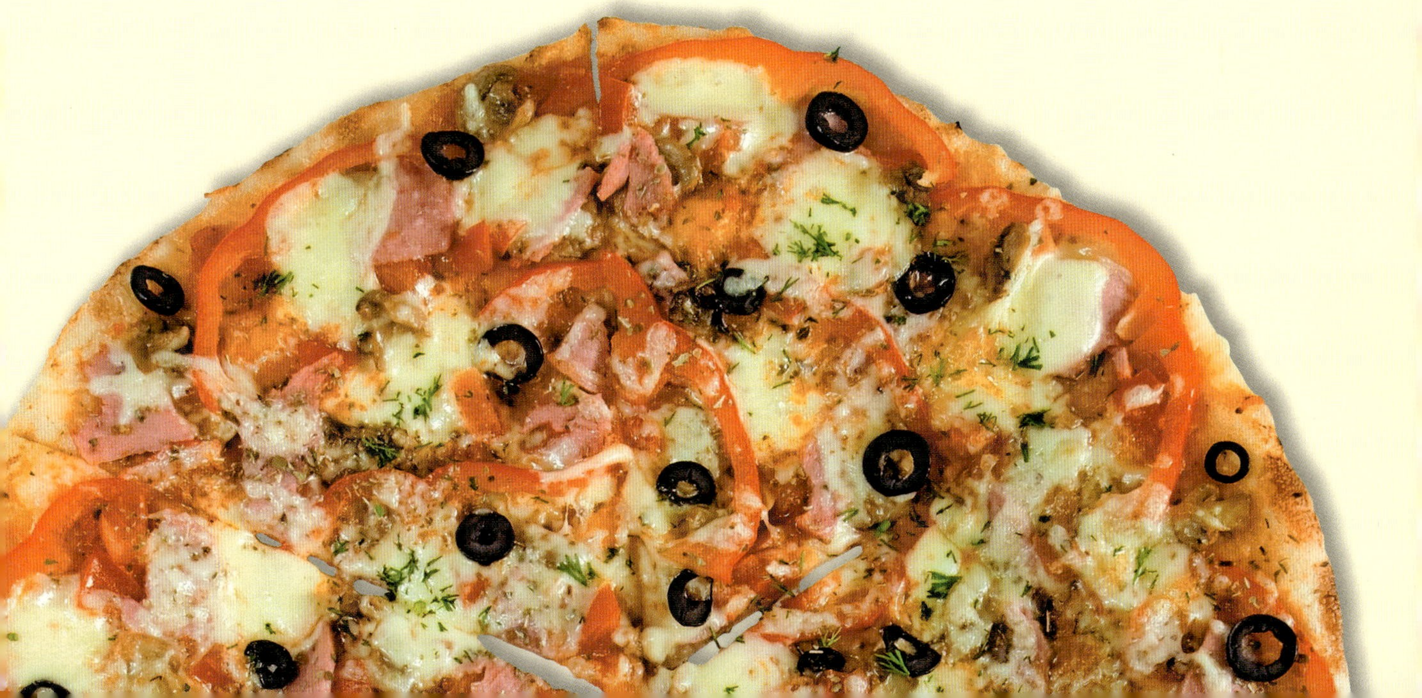

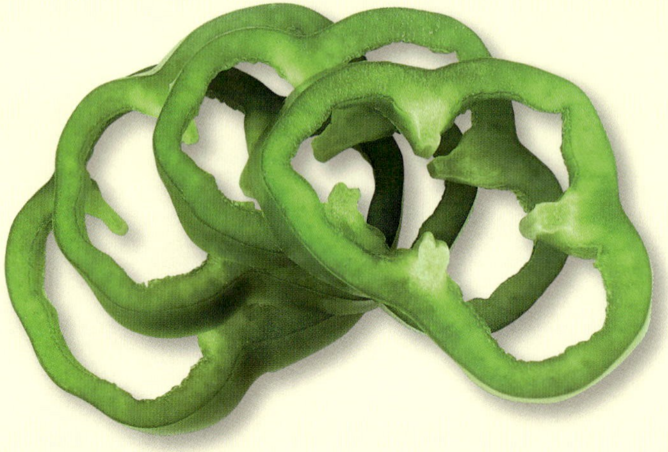
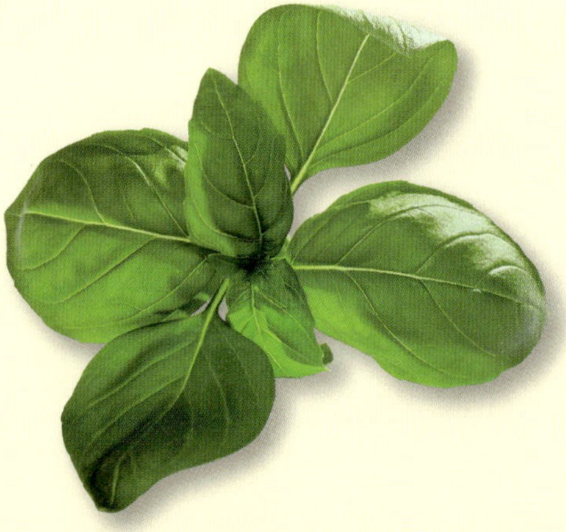
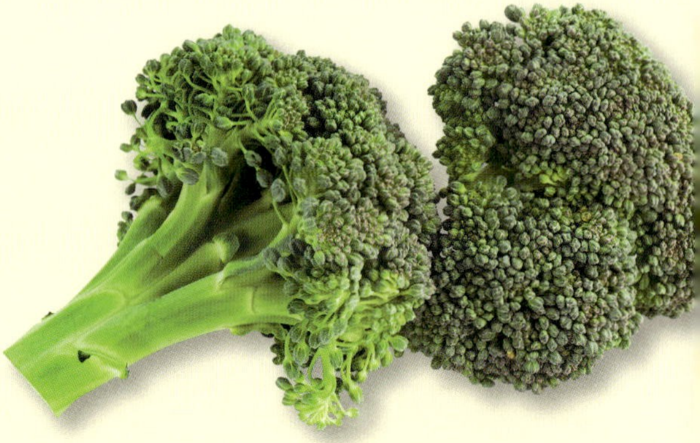

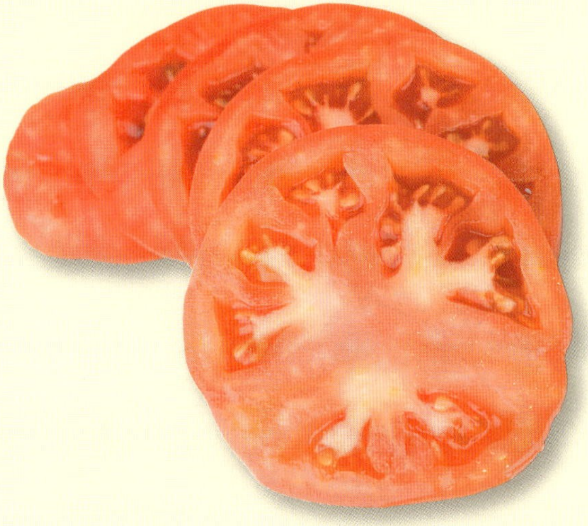

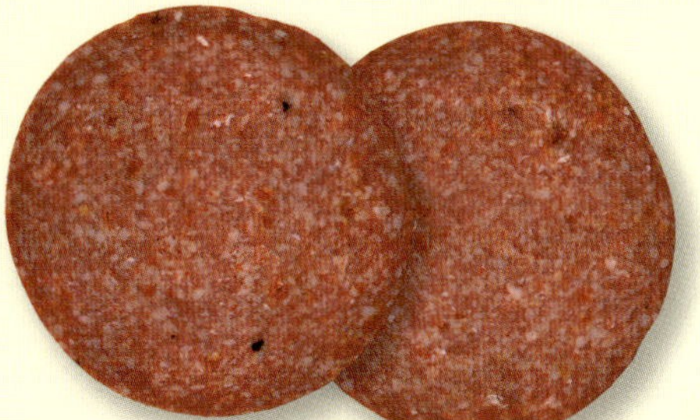

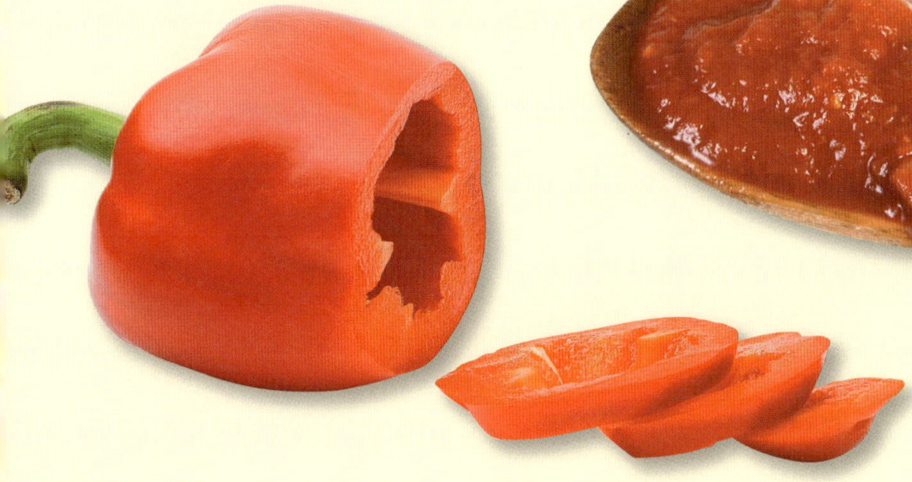

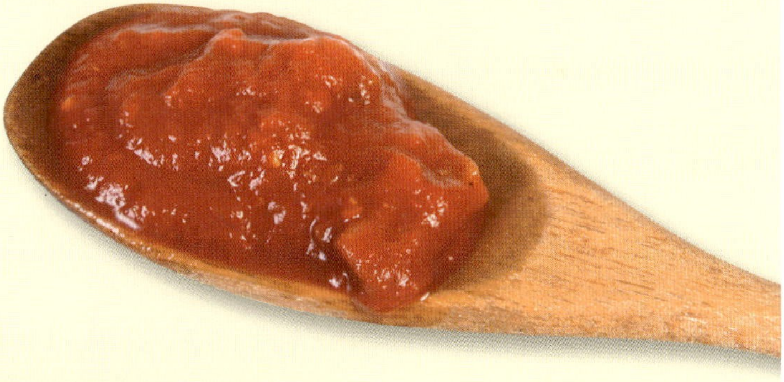

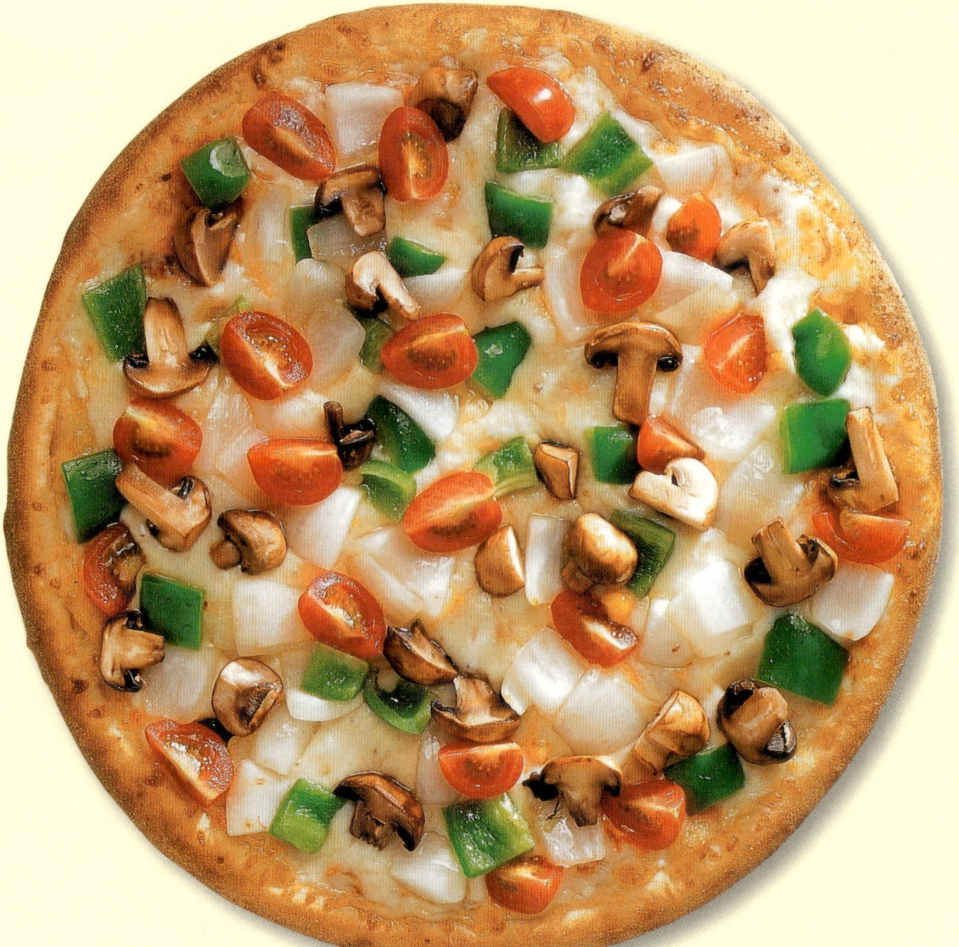

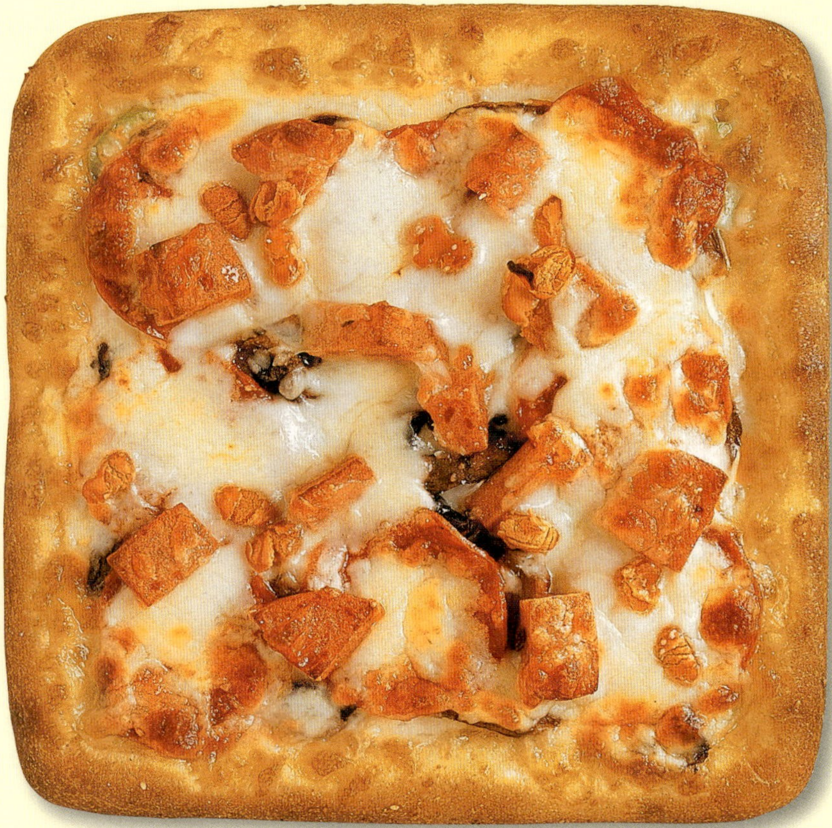

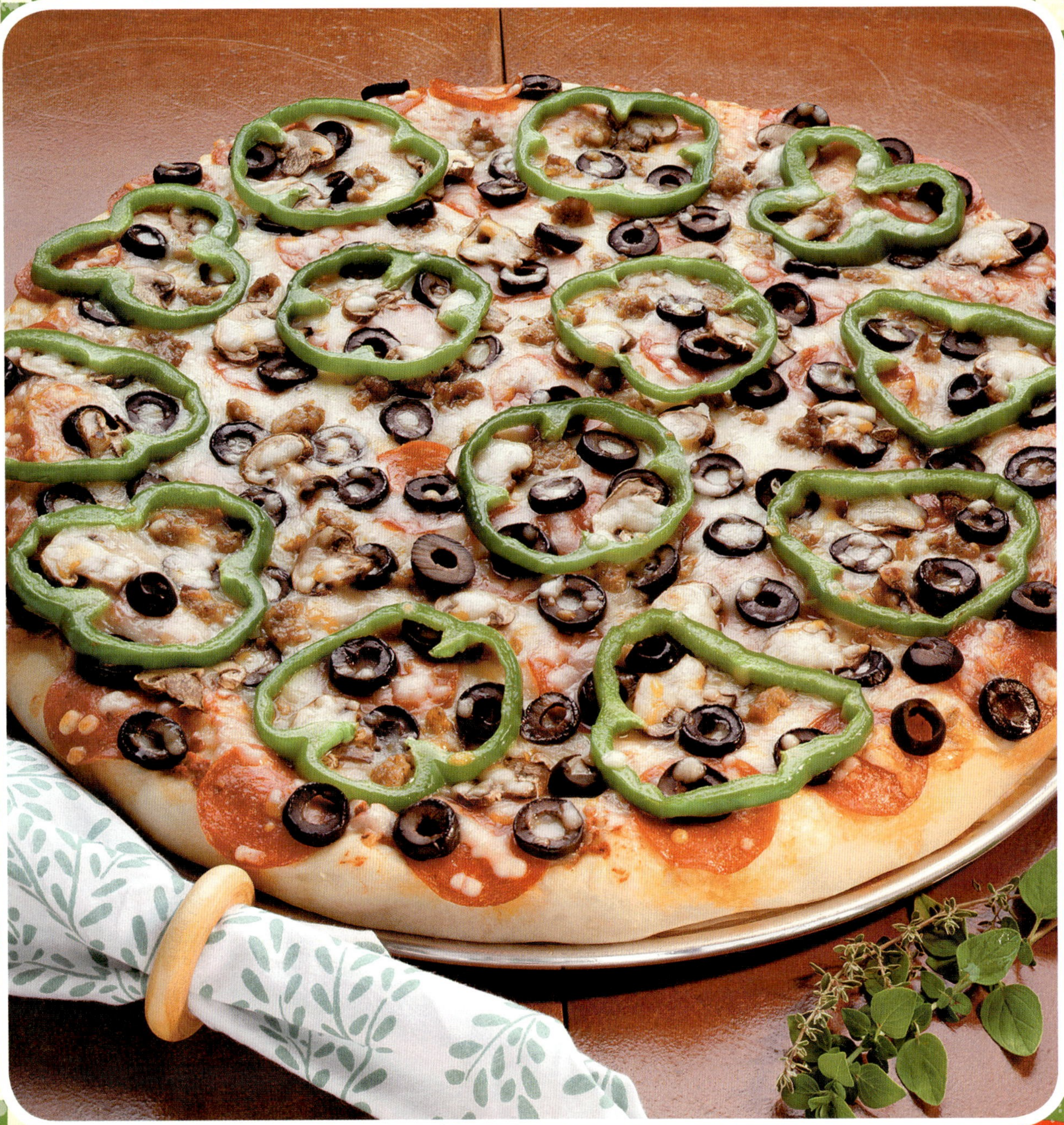

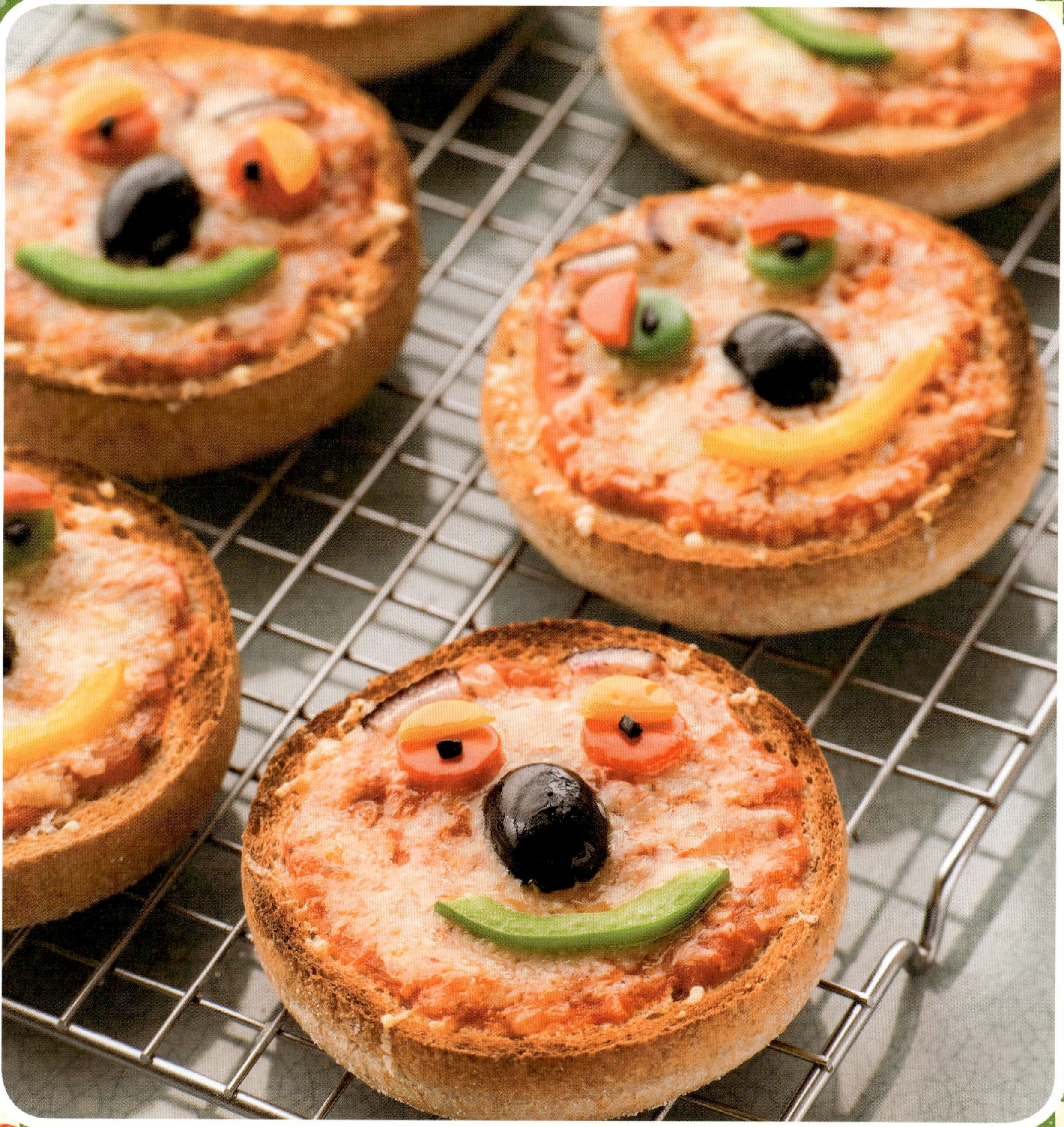

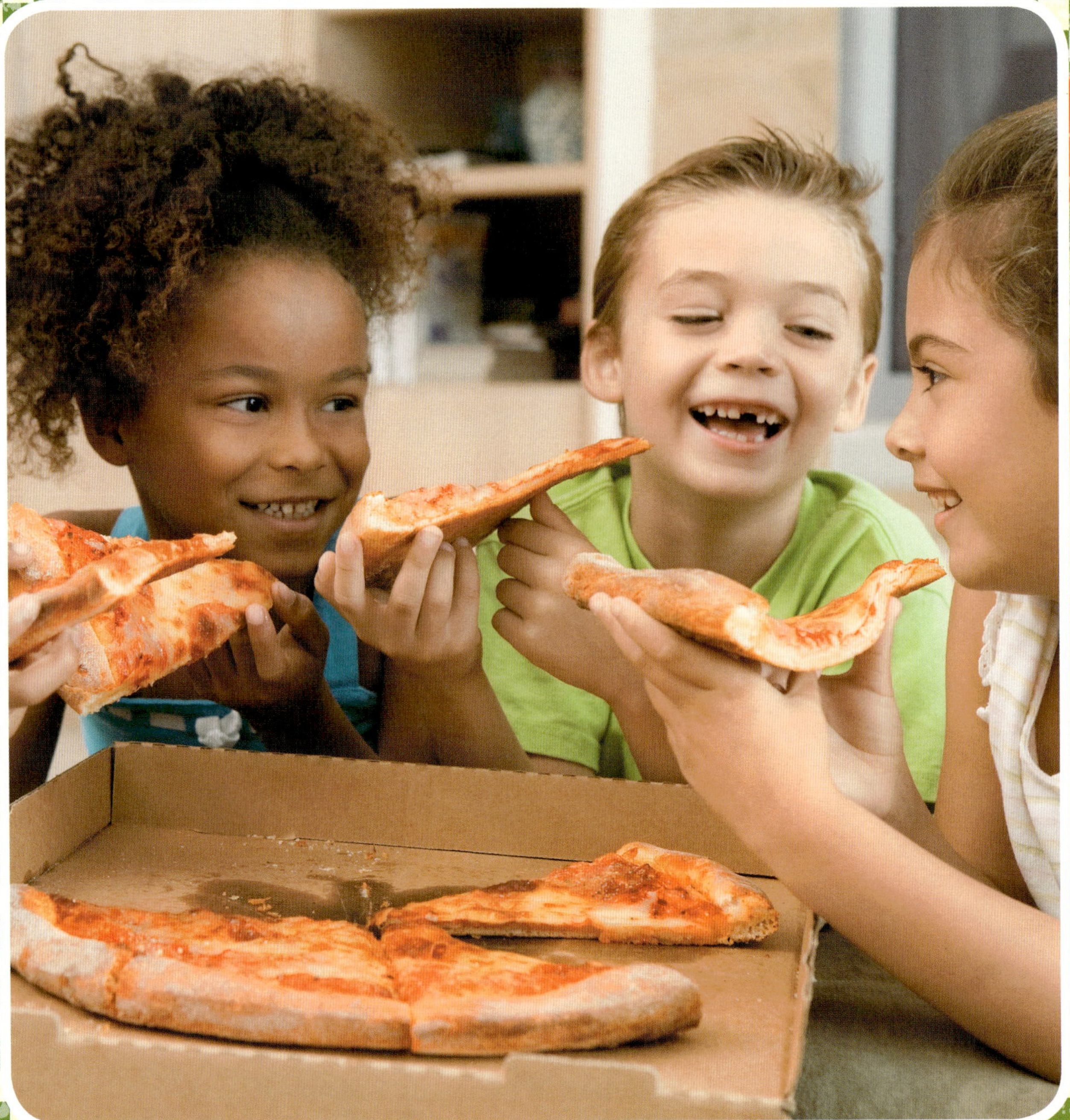